The Principal Farmer Has A Farm

By Patrick S. Muhammad

www.PSMEnterprise.com
info@psmenterprises.com
(404) 207-0544

www.RATHSIPUBLISHINGLLC.com
RATHSIPUBLISHING@GMAIL.COM

Copyright © 2016 Patrick S. Muhammad
Illustrations © 2016 AL Danso

ISBN: 978-1-936937-94-3

All Rights Reserved. No part of this book may be reproduced, stored in retrieval systems, or transmitted in any form, by any means, including mechanical, electronic, photocopying recording, or otherwise without prior written permission of the author and publisher. Printed in The United States of America.

This book is dedicated to my wife, Ishtar Muhammad,
who believed in my vision so much that we left our comfortable subdivision lifestyle for the "farm life."
Her sacrifice will allow us to prepare a better future for our children, grandchildren, and generations to come.

To my children, Ishlah, Ishijah, & Ishstafah, who have accepted being those "weird" children who live on a farm.
Your hard work will pay off for each and everyone of you very soon.
Keep loving each other and paving the way for the future.

To my family and friends, thank you for always supporting my endeavors.
Special shout out to Wayne "Chip" Swanson, Herb Muhammad and the Moo Squad.

To The Honorable Minister Louis Farrakhan, Thank you for The Time & What Must Be Done- #36,
it changed my life Forever.

Here We Grow!!!

Its 4:45am and all through the house,
there is only one person moving, the farmer,
and he's as quiet as a mouse.
Out the back door without making a sound
he sneaks pass the cat who's creeping around.

Its almost dawn and he nears the barn
to catch Mr. Roo
who's about to take his first yawn,
COOK-A- DOO-D- DOO,
is what he hears next,
and with that sound other creatures begin to stretch.

All of a sudden the farmer takes a deep breath,
"Yo Cowwssss" is heard and to whom you say,
a herd of cows respond and "MOO" his way.
He checks them all and they love his call,
they always come running hoping a treats involved.

Now down to the coop to check on the eggs, soon mom will be waking for breakfast in bed.

But now to the kennels to see who's awake, its puppies and dogs all tails array, no fussing or fighting just excitement and glee all saying to the farmer, " Walk me, Walk me "

Last check for the morning and what do you see
already in action are the busy bees.
Back and forth to the garden for honey they go
and from their help the vegetables grow, grow, grow.

Now into the house to start what you my ask,
next, it's preparing for school is the Principal's task.

Students are arriving with glee in their eyes, each day their excited, knowing their teachers classrooms will be full of surprise.

Off to class no time to waste,
education is key in this knowledge race.

Lunch is cooking and stomachs are growling, "eat all your vegetables", to keep your principal smiling.

All loaded and seated the students obey,
the buses are rolling some teachers whisper hooray.

Its time to head home for the principal you see it's back to the farm to plant a tree.

The Principal and Farmer is tiring from his daily routine, but here come his children with homemade ice cream.

Tomorrow's a new day and he must start again,
For the Principal Farmer, work is his best friend.

For more information about Your Faith Farms and tours.
YourFaithFarms.com
yourfaithfarms@gmail.com
(404) 207-0544

www.ingramcontent.com/pod-product-compliance
Lightning Source LLC
Chambersburg PA
CBHW041633040426
42446CB00024B/3497